ROSEN *Verified*

NATURAL DISASTERS

HURRICANES

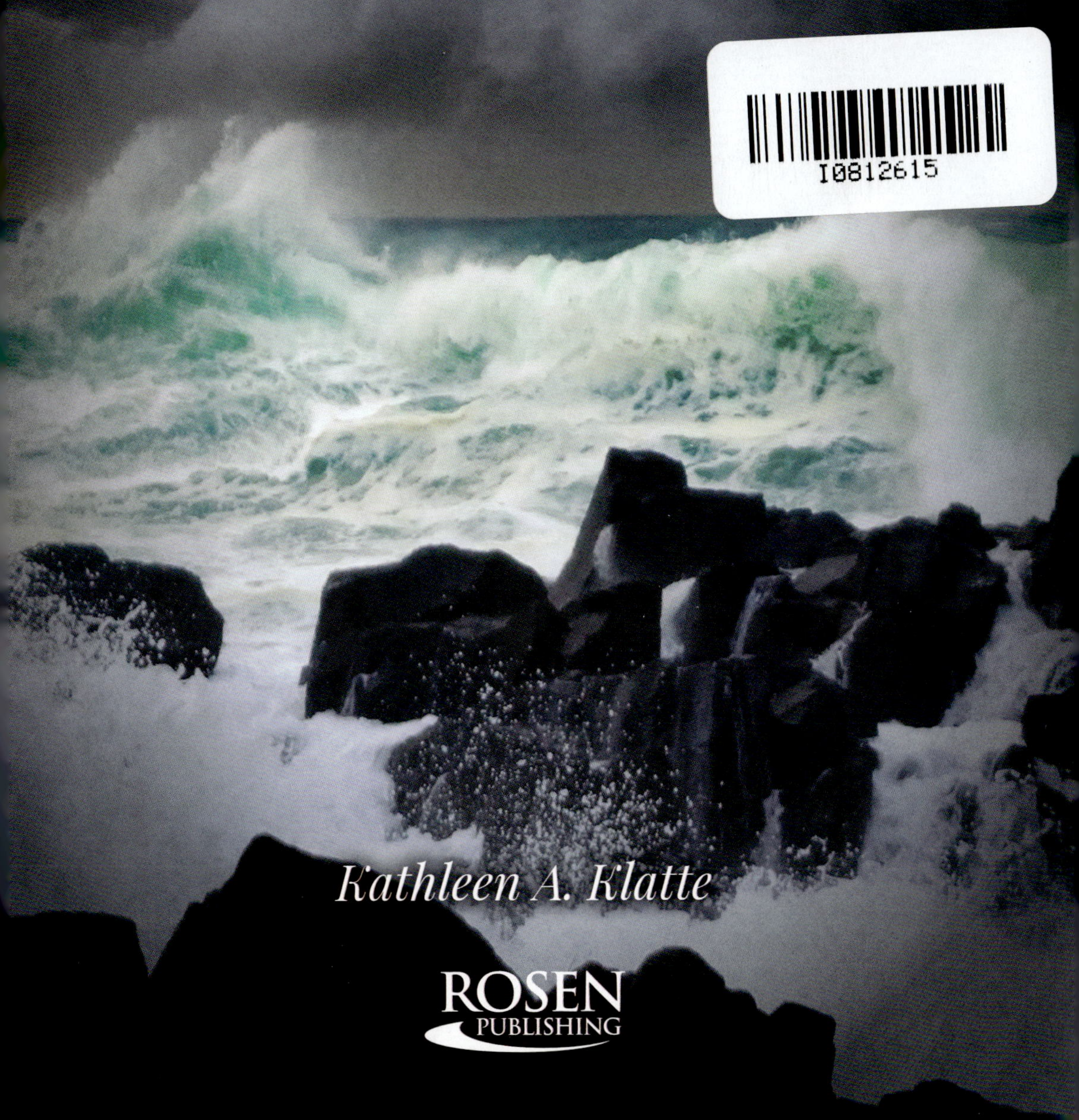

Kathleen A. Klatte

ROSEN PUBLISHING

Published in 2023 by The Rosen Publishing Group, Inc.
2544 Clinton Street, Buffalo, NY 14224

First Edition

Editor: Greg Roza
Book Design: Michael Flynn

Photo Credits: Cover Maximillian cabinet/Shutterstock.com; (series background) PinkPueblo/Shutterstock.com; p. 5 Vladi333/Shutterstock.com; p. 6 Zakhar Akulov/Shutterstock.com; p. 7 Fouad A. Saad/Shutterstock.com; p. 9 glenrichardphoto/Shutterstock.com; p. 11 (ocean and pier) Drew McArthur/Shutterstock.com; p. 11 (map) Yaska/Shutterstock.com; p. 12 Lynne Sladky/AP Photo; p. 13 Petr Merkurev/Shutterstock.com; p. 15 aappp/Shutterstock.com; p. 16 canadastock/Shutterstock.com; p. 17 Rainer Lesniewski/Shutterstock.com; p. 19 (waves) Paul Hampton/Shutterstock.com; p. 19 (hurricane scale) udaix/Shutterstock.com; p. 21 EHStockphoto/Shutterstock.com; p. 23 Skylines/Shutterstock.com; p. 25 MicheleMidnight/Shutterstock.com; p. 27 Cire notrevo/Shutterstock.com; p. 29 iKinoto/Shutterstock.com; pp. 31, 32 Joseph Sohm/Shutterstock.com; p. 33 https://commons.wikimedia.org/wiki/File:Andrew_1992-08-23_1231Z_(Cropped).png; p. 35 Nathan Holland/Shutterstock.com; p. 37 Brian Derr/Shutterstock.com; p. 39 EngineerPhotos/Shutterstock.com; pp. 40, 41, 43 Gemunu Amarasinghe/Shutterstock.com; p. 45 Eric Isselee/Shutterstock.com.

Library of Congress Cataloging-in-Publication Data

Names: Klatte, Kathleen A., author.
Title: Hurricanes / Kathleen A. Klatte.
Description: Buffalo : Rosen Publishing, [2023] | Series: Rosen verified: natural disasters | Includes index.
Identifiers: LCCN 2022027767 (print) | LCCN 2022027768 (ebook) | ISBN 9781499469554 (library binding) | ISBN 9781499469547 (paperback) | ISBN 9781499469561 (ebook)
Subjects: LCSH: Hurricanes--Juvenile literature.
Classification: LCC QC944.2 .K56 2023 (print) | LCC QC944.2 (ebook) | DDC 551.55/2--dc23/eng/20220630
LC record available at https://lccn.loc.gov/2022027767
LC ebook record available at https://lccn.loc.gov/2022027768

Manufactured in the United States of America

CONTENTS

MONSTER STORMS

What if a storm could move from the Caribbean all the way up the eastern coast of the United States? What if a storm could cause more than $8 billion in damage? What if stores in New York couldn't get deliveries because a storm flooded New Jersey? It's not the newest Hollywood blockbuster—it was Hurricane Irene in 2011.

If you live near a coast, you'll probably experience a hurricane at least once in your life. A hurricane is a type of tropical **cyclone** that forms over warm ocean water. Hurricanes are more powerful and more common closer to the **equator**. They're stronger over the ocean and weaker on land.

Hurricanes can cause terrible damage. Heavy rains cause floods. High winds can blow down trees and utility poles. Some hurricane winds are strong enough to blow the roofs off houses.

How can one storm be so big and cause so much damage? How can people prepare for something like that?

Meteorologists are scientists who study weather. They use **radar** and **satellites** to track storms. They develop ways to warn people when a dangerous storm is coming. They help local authorities make plans to keep people safe.

Average hurricanes are about 300 miles (483 km) across. They're so large that they can be seen from space. They're some of the strongest storms on Earth. They can cause huge amounts of damage.

SO MANY CYCLONES!

Hurricanes happen all over the planet, except for the equator and the poles. At the equator, there is no Coriolis force to create **rotation**. At the poles, the water is too cold.

Tropical cyclones have different names. In North America and the Caribbean, they're called hurricanes. In the Philippines, Japan, and China, they're called typhoons. In Australia, they're just called cyclones. That's a lot of different names for the same kind of storm!

Hurricane-force winds blow at the poles. However, the water is too cold to form a hurricane. This kind of storm is called a polar vortex.

THE CORIOLIS FORCE

The Coriolis force pushes air currents to the right in the Northern **Hemisphere** and to the left in the Southern Hemisphere. It's caused by Earth's rotation. It's what makes cyclones spin. It's weakest at the equator and strongest at the poles.

FAST FACT

The states that have had the most direct hits from hurricanes are Florida and Texas. Pennsylvania has had only one direct hit from a hurricane, in 1898.

WHAT MAKES IT A HURRICANE?

Blizzards have ice and snow. Tornadoes have funnel clouds. So what makes a hurricane?

A hurricane is a massive storm that forms over tropical ocean water. The Coriolis force creates rotation. This forms a cyclone. A storm is officially declared a hurricane when it has surface wind speeds greater than 74 miles (119 km) per hour.

A storm can take two or three weeks to develop into a hurricane. The storm can move over a great distance before it breaks up.

LATITUDE AND LONGITUDE

Latitude is the distance north or south of the equator. Longitude is the distance east or west from the prime meridian. That's an imaginary line that runs from the North Pole to the South Pole. Both latitude and longitude are measured in degrees.

FAST FACT

A hurricane can't form within five degrees of the equator. The Coriolis force is too weak to create rotation in the storm.

There have been many ideas for stopping hurricanes. They range from using nuclear bombs to cooling tropical seas with icebergs. Hurricanes are simply too large and powerful for these plans to be practical.

HOW DOES IT WORK?

Warm ocean water powers a hurricane. It works sort of like a battery. As long as the hurricane has a supply of warm, moist air, it grows stronger. Once the storm is over land, it starts to weaken. This happens because the air over land is drier. Passing over mountains can also break up a storm.

This is why a hurricane that moves up the eastern coast of the United States might still be a hurricane when it hits New York. The storm has had ocean water to fuel it. A hurricane that comes ashore and moves inland will probably weaken much faster.

THE LAW OF AVERAGES

About ten tropical storms form in the Atlantic, Caribbean, and Gulf of Mexico each year. Six of those will probably develop into hurricanes.

PATH OF HURRICANE IDA

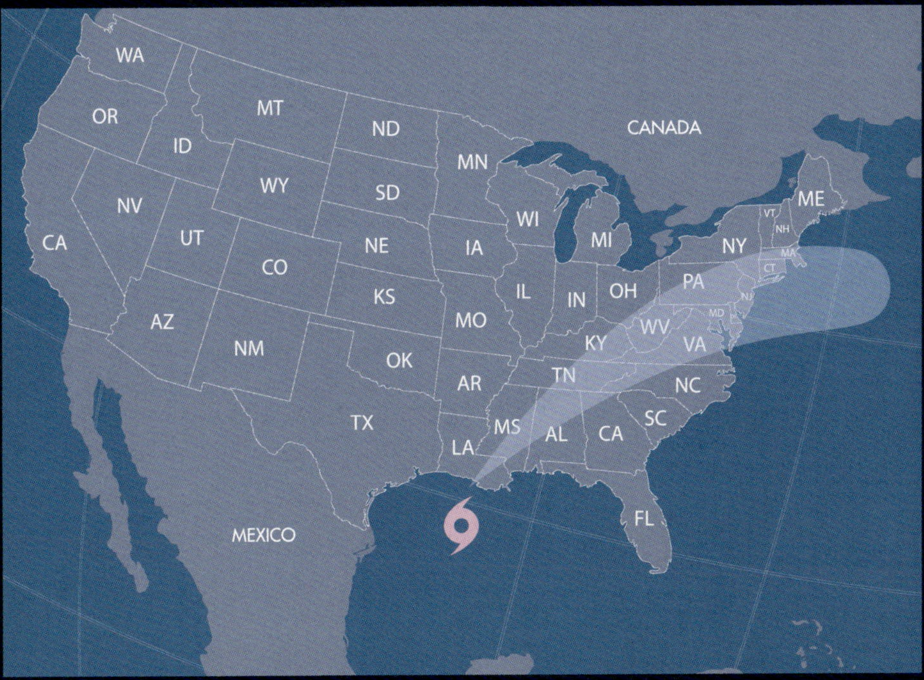

Hurricane Ida was one of the strongest hurricanes to strike Louisiana. The storm weakened quickly as it moved away from the coast. It still caused lots of flooding and damage.

STAGES OF A HURRICANE

A hurricane starts as a low-pressure system. This is a system that sucks in lots of air. Hurricanes form over water. The water must be above 79 degrees Fahrenheit (26 degrees Celsius). The water forms storm clouds. The clouds grow in a circular pattern.

A lot goes into the weather forecast you see on TV. Meteorologists collect data gathered from satellites and other sources. Computers process the data. Then they generate models of future weather patterns.

BAROMETRIC PRESSURE

A barometer is an instrument used by meteorologists. It measures changes in air pressure. Air pressure is the weight of all the layers of Earth's atmosphere. It's greatest at sea level.

If the wind speed is less than 39 miles (63 km) per hour, the storm is called a tropical depression.

When the wind speed becomes more than 39 miles (63 km) per hour, it becomes a tropical storm. Tropical storms are assigned names.

When the wind speed is greater than 74 miles (119 km) per hour, the storm becomes a hurricane.

ANATOMY OF A HURRICANE

The center of a hurricane is called the eye. It's very calm. The eye can be anywhere from 5 to 120 miles (8 to 193 km) across. Most are about 20 to 40 miles (32 to 64 km) across.

The eye is surrounded by the eyewall. This is a very dangerous part of the storm. It often contains the highest wind speeds.

The outer edges of the storm are called rainbands. They produce heavy rainfall. This is the part of the storm that causes floods.

FLASH FLOODS

A flash flood is one that occurs within six hours of heavy rainfall. Sometimes it can happen within three hours. Flash floods can often have strong currents. People shouldn't try to drive across flooded roads because their car can be swept away.

The eye of a hurricane is miles across. It's clearly visible from space. You can also see the circular structure of the storm.

ARE THERE HURRICANES IN SPACE?

So far, we haven't detected any other planets with warm oceans like Earth's. Without oceans, there can't be storms exactly like we have here on Earth. However, scientists think that gas giant planets have some type of cyclone.

HOW BIG, HOW FAST, HOW FAR?

Hurricanes are huge storms. Most hurricanes are about 300 miles (483 km) across. The eye of the storm can be 20 to 40 miles (32 to 64 km) across. Dangerous winds can extend hundreds of miles beyond the storm's center.

An average hurricane is about 300 miles (483 km) wide. That's slightly larger than the Grand Canyon!

FAST FACT

Did you know that some hurricanes spawn tornadoes? In 1964, a tornado formed by Hurricane Hilda killed 22 people in Louisiana.

Hurricane-force winds can extend over 100 miles (161 km) from the center of the storm. Tropical storm–force winds can extend up to 300 miles (483 km).

A storm is considered a hurricane when wind speeds are greater than 74 miles (119 km) per hour. These winds can damage the outside of houses and blow down tree branches. A storm is considered a major hurricane when wind speeds are greater than 110 miles (177 km) per hour. These winds can destroy houses and uproot trees.

HOW DO YOU MEASURE WIND?

Hurricanes are ranked by the Saffir-Simpson Hurricane Wind Scale. The scale is made up of five **categories**. These categories are based on wind speed and storm surge. Storms are ranked by how much death and damage they're likely to cause.

The weakest hurricanes are Category 1. These storms will cause flooding and some wind damage. The strongest are Category 5. These storms level houses and cause major flooding.

The scale was developed in the 1970s by an engineer named Herbert Saffir and a meteorologist named Robert Simpson.

The Saffir-Simpson scale helps people plan for storms. When a hurricane is coming, people might need to board up their windows. If the storm is strong enough, they might even need to **evacuate**.

STORM SURGE

A hurricane can cause sea levels to rise. This is called storm surge. It can cause the sea to rise 20 feet (6 m) above normal levels. This is a cause of major flooding.

SAFFIR-SIMPSON HURRICANE WIND SCALE

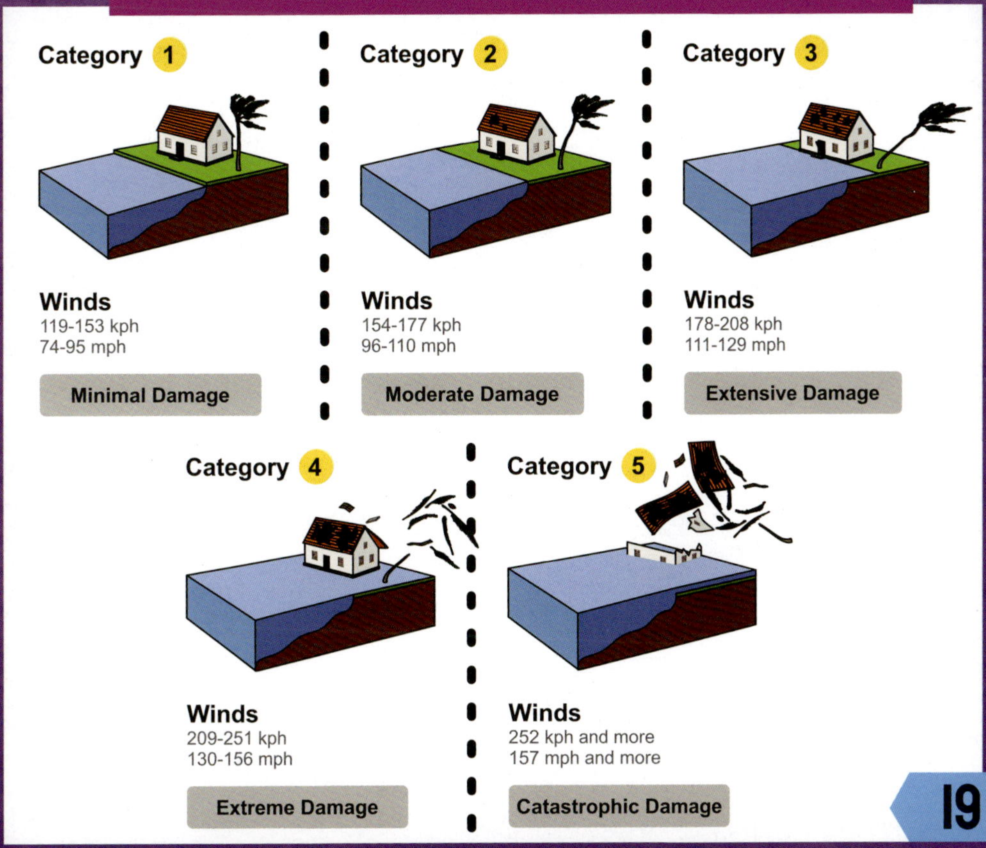

WHAT'S IN A NAME?

The National Hurricane Center (NHC) has named tropical storms and hurricanes since 1950. Names make it easier to share information about the storm. A storm is named if its wind speed is greater than 39 miles (63 km) per hour. It keeps the same name if it turns into a hurricane.

Names alternate between male and female. The names are used in alphabetical order. If there are more than 26 named storms in a season, the rest of the storms are named for letters of the Greek alphabet.

Some storms are so dangerous that their names are never used again.

WINTER STORMS

Only tropical storms and hurricanes are given official names. However, the Weather Channel names winter storms that are expected to be very large or dangerous.

FAST FACT

The word "hurricane" comes from Huracan or Hurakan. That was the name of a Mayan god of storms. Mayan beliefs spread throughout the Caribbean. Spanish sailors began to call the deadly storms "hurricanes."

Many agencies work together to gather information about storms. Having an official name makes it easier to share information. Can you imagine a TV meteorologist talking about a storm using a map position instead of a name?

WATCHES AND WARNINGS

The NHC issues hurricane watches and warnings. A hurricane watch is posted when hurricane-force winds are possible. Watches are posted 48 hours (2 days) in advance. A hurricane warning is posted when hurricane-force winds are expected. Warnings are posted 36 hours (1.5 days) in advance.

People who live in coastal areas know to take these announcements seriously. They check to be sure they have supplies for the storm. Older people might consider evacuating before an official order is given.

EVACUATION ORDERS

Evacuation orders are issued by government agencies when there is imminent public danger. It means that the area is very unsafe. If people stay, they risk serious injury. Rescue workers may not be able to reach them.

National Hurricane Preparedness Week is usually the first week of May. This is to give people time to assess their risk and prepare for hurricane season.

VERIFIED

The National Oceanic and Atmospheric Administration (NOAA) is part of the U.S. Department of Commerce. NOAA is the agency that monitors weather and climate. You can learn more about it at **www.noaa.gov**.

LIVING THROUGH A HURRICANE

Some coastal areas experience a hurricane or tropical storm almost every year. People who live there know how to prepare for storms. They have battery-powered radios, flashlights, and plenty of bottled water. They listen carefully to watches and warnings. They know their evacuation route.

Some companies specialize in building "hurricane-proof" houses. They're round to withstand the wind better. No matter how sturdy your house is, it's important to follow evacuation orders from the authorities.

HURRICANE SEASON

The Atlantic hurricane season is June 1 to November 30. The Eastern Pacific hurricane season is May 15 to November 30.

People who live where hurricanes are common board up their windows to protect them. They put away lawn furniture so it doesn't blow around. They use sandbags to protect doorways.

STORM SUPPLIES

- Flashlights
- Radio
- Batteries
- Bottled water
- Pet carrier & supplies
- Phones & chargers
- Printed list of important phone numbers
- Cash
- A full tank of gas/charged battery in the car

AFTER THE STORM

After a hurricane passes, there's a big mess to clean up. If you're lucky, you might just have to pick up some branches and put things back outside. If the storm was stronger, your family might need to remove boards from windows or pump out the basement.

Sometimes it takes days or even weeks to recover from a hurricane. Fallen trees can block roads. Utility poles can be damaged. Electricity and phone service might be out for a while.

STAYING SAFE

Sometimes people die after a storm has passed. Often this is because of power outages. Candles can cause fires. Generators and camp stoves are meant to be used outdoors. They can cause carbon monoxide poisoning if there isn't enough ventilation, or air flow.

VERIFIED

The American Red Cross is a **humanitarian** organization. It responds to natural disasters like hurricanes. It brings supplies and aid to places that need help. Find out more about it at **www.redcross.org**.

If you evacuate during a storm, it's important to wait until the authorities say it's safe to return home. Roads might need to be cleared. Utility companies have to make sure the electricity is safe.

HISTORIC STORMS

Some of the worst hurricanes predate modern naming. The Galveston Hurricane of 1900 is considered the worst weather disaster in U.S. history. The storm caused 8,000 deaths and $30 million in property damage.

The Great Atlantic Hurricane of 1944 made landfall in Long Island as a Category 3 hurricane. It sank five ships, including a navy destroyer.

Hurricane Camille in 1969 is the second most intense hurricane to make landfall in the United States. The storm destroyed measurement equipment when it struck. Its exact wind speed is unknown.

FAST FACT

When Hurricane Ivan damaged the Alabama Gulf Coast Zoo in 2004, an alligator named Chucky escaped. He was captured by law enforcement and a team from Florida's Gatorland park. Chucky was safely returned to the zoo.

VERIFIED

The Federal Emergency Management Agency (FEMA) is the federal organization responsible for assisting the public after a major disaster. It also offers information on how to prepare for emergencies. Find out more at **www.fema.gov**.

Guinness World Records lists Hurricane Wilma as the most intense hurricane recorded by the National Hurricane Center. Wilma was the last major hurricane of 2005.

THE PERFECT STORM (1991)

The Perfect Storm, or Halloween Storm, is best known for sinking the fishing boat *Andrea Gail*. The events of the storm are told in the book and film *The Perfect Storm*.

The Perfect Storm was a **nor'easter** that combined with Hurricane Grace. The massive storm caused flooding from North Carolina to the Canadian Maritimes. New England suffered damage from high winds.

The storm dwindled, then reformed as a hurricane at sea. The second hurricane was not named.

LINDA GREENLAW

Linda Greenlaw was the only female swordfishing captain on the East Coast at the time of the Perfect Storm. She's best known as the captain of the *Hannah Boden*, the sister ship of the *Andrea Gail*. Greenlaw, now a best-selling author, was the last person to have contact with the *Andrea Gail*.

FACT OR FICTION?

The Perfect Storm was the basis for a book by Sebastian Junger. It was later made into a movie. The film is a **dramatization** of actual events. Some things were exaggerated for the movie. Others are **speculation**.

The *Andrea Gail* never sent a distress call. Its emergency beacon washed up on Sable Island a few days after the storm. Sable Island is off the coast of Nova Scotia in Canada. The exact fate of the boat and its crew remains a mystery.

HURRICANE ANDREW (1992)

Hurricane Andrew was one of the worst hurricanes to strike the United States. It struck south Florida as a Category 4 hurricane. Winds of 142 miles (229 km) per hour were measured before the equipment was destroyed by the storm. More than 25,000 homes were destroyed.

FAST FACT

An urban legend says the spires of Cinderella Castle in Disney World can be removed in case of a hurricane. According to Disney, the rumor is false. The castle is built to withstand the strongest storms. So are the other buildings in Disney World. The park's hotels are some of the safest buildings to ride out a major storm.

The energy released by the eye and eyewall of Hurricane Andrew was 5,000 times the output of a nuclear power plant.

The storm crossed the Gulf of Mexico and struck Louisiana as a Category 3 hurricane. It created a tornado and 8-foot (2.4 m) storm tides.

Hurricane Andrew caused $26.5 billion of damage in the United States. There was an additional $250 million in damage in the Bahamas. There were 23 deaths in the United States and 3 in the Bahamas.

HURRICANE KATRINA (2005)

Hurricane Katrina was the deadliest hurricane to strike the United States since 1928. The storm struck Florida, then crossed the Gulf of Mexico. It made landfall in Louisiana as a Category 3 hurricane.

The storm surge was 10 to 20 feet (3 to 6 m) above normal levels. The flooding breached the **levees** that protect the city of New Orleans. Flood water was still being pumped out of the city six weeks after the storm.

Hurricane Katrina is currently considered the costliest natural disaster in U.S. history. The storm caused $100 billion in damage. It claimed at least 1,200 lives.

THE LITTLE ZOO THAT COULD

The Alabama Gulf Coast Zoo was the first zoo in the United States to evacuate its animals from the path of a hurricane. The park was the subject of an Animal Planet TV series. The show documented the zoo's recovery from Hurricane Ivan. The zoo was also damaged by Dennis and Katrina in 2005.

Flood waters are a major hazard after a hurricane. Water can be **contaminated** with sewage, chemicals, or dead animals. It can also conceal dangerous objects. Places that have been flooded can develop mold that makes people sick.

SUPERSTORM SANDY (2012)

Hurricane Sandy caused immense destruction in the Caribbean. Then the storm moved up the eastern coast of the United States. It mixed with a cold front to form a massive "superstorm."

Eight million people lost power from the storm. Flooding and power outages in New Jersey caused long lines at gas stations. Flooding in New York City closed subways. A full moon caused higher than usual tides, in addition to the storm surge.

The storm caused $70 billion in damage. It was responsible for almost 200 deaths.

CLOSING DOWN WALL STREET

Superstorm Sandy shut down Wall Street for the first time since the 9/11 attacks. It was the first weather-related closure since Hurricane Gloria in 1985. Wall Street was closed for two days. This was the longest weather-related closure since 1888!

FAST FACT

Most people who die during hurricanes drown. Storm surges can cause water levels to rise very quickly. Roads can be washed out. Flash floods can have very strong currents.

People don't expect a storm like Sandy to come ashore so far north. Places like New York City don't have good plans for dealing with this kind of flooding.

PREDICTING HURRICANES

Scientists use many tools to predict hurricanes. They monitor the atmosphere with radar and satellites. Planes and ships measure temperature and wind speed. Weather buoys monitor the seas.

All this information is **transmitted** to labs. Meteorologists put it all together, like pieces of a puzzle. They look at records of past storms to get ideas of what could happen. They use computers to generate models of the storm's potential path.

WHAT KIND OF DATA?

- Air temperature
- Water temperature
- Barometric pressure
- Tides
- Currents
- Other local storm systems
- Time of year
- Wind speed and direction

TIDES

Tides are the regular rising and falling of the sea. They're caused by the gravitational pull of the moon and sun. During a new moon or full moon, the tides are higher. Higher tides that occur during a severe storm can cause flooding.

A weather buoy is a small, unmanned floating object. It contains equipment to track and transmit things like air temperature and wind speed.

TRACKING HURRICANES

Once a tropical storm system forms, the National Hurricane Center tracks it. It posts official updates every six hours.

Hurricanes are very large and dangerous. It's important to know where the storm is going and how strong it is. Meteorologists work very hard to provide accurate forecasts. They want to give people as much time as possible to prepare for the storm or evacuate.

VERIFIED

The National Hurricane Center (NHC) is part of NOAA. It monitors all tropical cyclones that form in the Atlantic or eastern Pacific Oceans. It posts public announcements for as long as the storm lasts. It issues official watches and warnings. You can learn more about it at **www.nhc.noaa.gov**.

NOAA and the U.S. Air Force have planes that fly into and around hurricanes. They carry **sophisticated** gear to take readings from the storm.

Sometimes storms weaken, then gain strength again. Sometimes the same storm makes landfall in two different places. Hurricanes are tracked closely until they disappear completely.

WHO STUDIES HURRICANES?

Scientists study hurricanes from different angles to try and understand how they work.

NASA (National Aeronautics and Space Administration) studies storms from space. Satellites can view storm systems from above. There are even instruments, or tools, on the International Space Station that measure wind speeds over the oceans.

Scientists at Woods Hole Oceanographic Institution (WHOI) study hurricanes from the water. They place instruments in the oceans to measure water temperature. This helps them understand all the factors that make up a hurricane.

VERIFIED

For over 90 years, the Woods Hole Oceanographic Institution has been a leader in marine science. It studies all aspects of the oceans—how they work, what lives in them, and human impact. You can learn more about it at **www.whoi.edu**.

Scientists study the environment many different ways. They take lots of different measurements. Then they combine their **observations**. They use computer programs to help figure out how hurricanes work.

LEARNING TO LIVE WITH HURRICANES

Hurricanes are the strongest storms on Earth. There's nothing that can be done to weaken a hurricane or alter its course. The best we can do is to learn as much as we can about them.

The more scientists know about how hurricanes form, the better they're able to predict them. Early warning is the best defense we have against hurricanes. The more warning we have, the better we are able to protect our homes. Early warning also helps authorities evacuate people in an orderly way.

Global warming is a factor in hurricanes. Earth naturally goes through cycles of heating up and cooling down. Human activity has generated lots of pollution. In the last 50 years, this has made global warming much worse. Rising temperatures make hurricanes more likely. Higher sea levels make flooding worse.

Even after the terrible destruction of Hurricane Katrina, people returned to their homes. They rebuilt. Scientists and engineers learned from the storm. They can create better safety features for the future.

Reducing greenhouse gas **emissions** won't make hurricanes go away, but it helps our environment. The more we know about how our world works, the better we'll be able to cope with natural disasters like hurricanes.

GLOSSARY

category: A group of people or things that are similar in some way.

contaminate: To make (something) dangerous, dirty, or impure by adding something harmful or undesirable to it.

cyclone: An extremely large, powerful, and destructive storm with very high winds that turn around an area of low pressure.

dramatization: A play, movie, or television show based on a work of writing or a real-world event.

emission: Something that is sent out (such as energy or gas) from a source.

equator: An imaginary circle around the middle of Earth that is the same distance from the North Pole and the South Pole.

evacuate: To leave (a dangerous place).

hemisphere: One half of Earth.

humanitarian: Concerned with making other people's lives better.

levee: A long wall of soil built along a river to prevent flooding.

nor'easter: A storm with winds that blow from the northeast.

observation: The act of carefully watching and listening in order to get information.

radar: A device that sends out radio waves for finding out the position and speed of a moving object (such as an airplane).

rotation: The act or process of moving or turning around a central point.

satellite: A machine that is sent into space and moves around Earth, the moon, the sun, or a planet.

sophisticated: Very complicated.

speculation: Ideas or guesses about something that is not known.

transmit: To send (information, sound, etc.) in the form of electrical signals to a radio, television, computer, etc.

INDEX